Jacob Vradenberg Brower, Society Minnesota Historical

Prehistoric Man at the Headwaters of the Mississippi River

Jacob Vradenberg Brower, Society Minnesota Historical

Prehistoric Man at the Headwaters of the Mississippi River

ISBN/EAN: 9783744791243

Printed in Europe, USA, Canada, Australia, Japan

Cover: Foto ©berggeist007 / pixelio.de

More available books at **www.hansebooks.com**

PREHISTORIC MAN AT THE HEADWATERS OF THE MISSISSIPPI RIVER.*

BY HON. J. V. BROWER.

I.

PRELIMINARY REFERENCES.

At pages 123 and 124, Vol. VII., Minnesota Historical Collections, prepared and submitted by me in 1889-93, the following appears:

Concerning the presumable fact, that, antedating the first known visit of white men at Lac La Biche, French voyageurs may have reached the basin, no reliable statement in writing is known to exist describing such visit. In the absence of any known record as to the movements of the French fur traders and voyageurs who first established themselves in lines of trade and traffic with the Indians across the northern portion of the territory which now constitutes the State of Minnesota, no definite record can be found concerning a mere probability that they may have reached Elk lake. To the writers of the future must be left the task of discovering the record of the manner in which "Lac La Biche" first became known to the French and of any visits they may have made to the locality, if any such record exists, which now seems doubtful. Certain it is that Mr. Morrison's letter is the only record of the first visit to the source of the Mississippi of which we have any knowledge.

Upon page 16 of my report to his Excellency the Governor of Minnesota, for the two years ending Dec. 1, 1894, the following tabulated historical record of the descent of title by possession appears:

*Abridged extract from the Journal of the Manchester (Eng.) Geographical Society, vol. XI., pp. 1-80, 1895; to which is appended an addendum, relating to the early visits of Mr. Julius Chambers and Rev. J. A. Gilfillan to Itasca lake, prepared for the Minnesota Historical Society by Mr. Brower.

Briefly stated, the actual possession of the Itasca basin may be approximately given as follows:

Preglacial ages.................... Possibly palæolithic man.
The Glacial period................ Possibly an Esquimaux occupancy.
Postglacial period................. The Mound-Builders.
The succeeding occupancy......... The Sioux Indians.
The Columbian period............. The Spanish.
Post-Columbian period............. The French and English.
The seventeenth or eighteenth cen-
   tury .... The Ojibway Indians.
The eighteenth century............ The Federal Republic.
Feb. 22, 1855..................... Ceded, by treaty between the United
                                    States and the Ojibway Indians.
1876-1891 .... Surveyed by the government and
                                    opened to pioneer settlement;
                                    Peter Turnbull and family and
                                    others.
1891 .... Set apart by law and dedicated as
                                    a public park forever.

Before describing the manner in which a recent discovery of the unmistakable remains of an extinct village of Mound-Builders was made near the geographical center of North America, a few preliminary references may be presented.

The writer disclaims any special or exhaustive knowledge in the field of archæological research, and presents the results of a very interesting and instructive discovery from the standpoint of a general desire to formulate ascertained facts for the benefit of those who cherish the advancement of scientific knowledge. It is impossible for the meditative explorer, grasping after a larger and more extended information, not to consider, so far as visible indications will permit, the existence, appearance, condition and habits of a people long since extinct, with whose relics, remnants, shell heaps, workshops, mounds, pottery and remains he has been brought into immediate contact while prosecuting geographical explorations.

The years gone by placed me between two fiercely contending tribes of North American Indians, at savage and bloodthirsty warfare, when tomahawks at the belt, paint, feathers and the scalplock braided from the top of the head constituted important preliminaries to the fierce struggles between wily warriors of the Sioux and Ojibway races.

Later on, during the Sioux outbreak of 1862 and the Indian war that followed, with my companions-at-arms, when we met the fierce Dakotas face to face, for supremacy or extermination,[1] the actual observance and participation then had accentuates an opinion now entertained, that probably the prehistoric race of men who occupied the upper waters of the Mississippi river basin were not extraordinarily different from the nations and tribes now receding before the enlightened encroachments of the English-speaking people. Time has brought its exorbitant and remarkable changes, and, making due allowance for the doubts engendered by the lapse of past ages, the cautious explorer, with some knowledge of aboriginal tribes, can intelligently study the relics and remains described in the following pages, written only because we now first certainly know that prehistoric man penetrated the wilderness of North America to the limit of the great continental watershed whence flow the precipitated waters, returning to the Atlantic through the delta of the Mississippi, the Gulf of the St. Lawrence and the Bay of Hudson; debouchures separated by distances not comparable with any other of a like hydrographic reference or geographical importance in the western hemisphere, if, indeed, in the whole world.

II.

THE DICKENSON MOUNDS.

Archæologists and historians have quite fully made known the existence of man in the Valley of the Mississippi at a very early and unknown date.

Mounds situated at St. Paul, Minn., and along many branches of the Upper Mississippi have long been known and fully described. Prof. T. H. Lewis has prosecuted investigations farther to the northward in the valley of the main river than perhaps any other professional archæologist. In all these well-known writings I fail to find any mention or description of the Dickenson mounds or earthworks, situated at

[1]General Sibley's expedition in 1863 from Fort Ridgely, on the Minnesota river, to Fort Abercrombie (recently abandoned), on the Red River of the North, about thirty miles south from the city of Fargo, N. D., thence to the Missouri river, where the city of Bismarck, capital of North Dakota, is now located. This expedition drove the Sioux across the Missouri, above the mouth of Apple creek, one of the results of the Sioux warfare against the white inhabitants of Minnesota in 1862.

Park Rapids, Minn., about twenty miles southeastwardly from Itasca lake. I had the pleasure of excavating a mound of this group in July, 1894, with the assistance of Dr. P. D. Winship. The unmistakable signs of the hand of man are visible in their construction, but our cursory examination did not develop nor determine their true character unless they were one time a work for defense or for a place of burial, and a further and more extended examination is necessary to determine the true origin of these twenty-two mounds of different sizes and heights. Persistent denudation with the plough and harrow upon Mr. Dickenson's farm has made little progress toward an intended annihilation of this group. We were unable to discover any prehistoric relics or the remains of the dead.

III.

A VOYAGE OF DISCOVERY DOWN THE MISSISSIPPI RIVER
FROM ITASCA LAKE.

At the time the Dickenson mounds were examined arrangements had been completed with Rev. T. M. Shanafelt, D. D., of South Dakota, and Rev. S. Hall Young of Iowa, for a voyage from Itasca lake to St. Paul, down the channel of the Mississippi, a distance of nearly six hundred miles. We remained at Itasca lake during the first week of August, in the full enjoyment of the superb and picturesque landscape scenery, then portaged six miles to the northward by team to avoid the Ka-ka-bi-kons rapids,[2] and launching our klinker-built boat, with a commissary supply in a convenient lighter, we sped on our way with unvarying success through the magnificent scenery, camping at the bluffs and enjoying the evergreens, extensive savannas[3] and tributary streams and lakes, until we encamped a week later at the remote and picturesque Lake Bemidji, called by the native Ojibway, Bem-e-jig-u-mag, meaning "The current of the river crosses the lake." Here we were suddenly made aware of geographical subdivisions of the upper basin of the Mississippi, for out from Bemidji lake the waters of the river soon plunge over a series of rapids,

---

[2] The first rapids on the Mississippi river north of Itasca lake, five or six miles distant by the channel of the river.
[3] Remarkable meadows of grass.

formed by glacial boulders, for a distance of twelve miles. So we know now, in addition to our Itasca basin, we also have a Bemidji basin, and lower down the main river, the Winnibigoshish and Cass lake basin, also the basin above the Pokegama;[4] these last two distinct basins constituting a division at the Winnibigoshish reservoir dam,[5] an artificial separation of the waters of that region. Speedily approaching the magnificent and island-dotted body of water named after Gen. Lewis Cass, the statesman and ambassador, we encamped at the old Northwest company trading post of a hundred years ago, at the mouth of Turtle river. There is not a vestige of this old post left, save only decaying emblems of disappearing and abandoned graves. That was a proper place for a painted and inscribed historical tablet, and we made it of oak.

A beautiful body of water next west from Cass lake, through the south end of which the Mississippi takes its course, I have designated Lake Elliott Coues, in honor of the distinguished gentleman whose scientific labors, with others, gave to the world the "Century Dictionary." Meeting him on the turbulent waters of Winnibigoshish lake, he stemmed the currents of the rapids and river to its source, and I gave his name to the lake, a simple tribute to persistent labors in the field of scientific research. Then we camped for a Sabbath rest at the Winnibigoshish reservoir dam, after having made numerous and friendly acquaintances among the different bands of Ojibway Indians residing along the course of our voyage.

In plate IV. this upper drainage basin of the Mississippi river is mapped, with designations of the localities where mounds, village sites, and other indications of the Mound-Builders have been found.

---

[4]The Falls of Pokegama, where the river plunges over a ledge of rock in place, as it issues out from the upper or headwater basin of the Mississippi which has been geographically known for many years and which contains about five thousand square miles and a thousand rivers and lakes of different sizes. This extensive headwater basin is hydrographically divided in the manner stated in the text. Leech lake on the south and Turtle lake on the north side of the main basin contribute a perennial flowage through streams bearing the same designations as the two lakes named.

[5]Constructed by the government of the United States for a reserve supply of water during the navigable season of the year, from St. Paul to the mouth of the Mississippi. This dam, and other similar ones in remote localities, are used to hold back precipitated waters in artificial reservoirs until about the 1st of August of each year, when the gates are raised and the flood of water replenishes the navigable channel of the river several feet in depth for hundreds of miles.

PLATE IV.

A Map
of the
Upper drainage Basin
of the
Mississippi River
above pokegama Falls.
Pre-historic Mounds,
Village sites and Indications
are marked ⬤

Scale in miles

IV.

THE MOUND-BUILDERS' HOME AT WINNIBIGOSHISH NEAR THE CUT FOOT SIOUX AND AT ITASCA LAKE.

Near our camp at the Winnibigoshish dam the ancient landmarks of prehistoric man appeared in abundance. Their burial mounds are situated on either side of the Mississippi at the lower extremity of the lake, which is near the peculiar waters of the Cut Foot Sioux, from which, by a portage of scarcely one mile, the waters of the Bow String, at the head of the Big Fork river, are reached. These last named waters flow to the Lake of the Woods, and, as it is known that pre-historic evidences of man exist on the banks of the Big Fork river, it is reasonable to presume that the portage from the Cut Foot Sioux to the Bow String and Big Fork was known and probably discovered by the Mound-Builders centuries before the aggressive Ojibway drove out the Sioux. The death and scalp of a Sioux warrior, killed in battle, who had lost a part of both feet, gave the unique name of "Cut Foot Sioux." A careful examination near the Winnibigoshish mounds brought to light the evidences of a former village site in that immediate neighborhood. The construction of the reservoir dam by the United States government had nearly obliterated one large mound, and the remaining sands, without great effort, gave forth very interesting translucent jasper and quartz spear-heads, arrow-points, mouldering skulls and bones, and quite a double handful of human teeth in a fair state of preservation could have been collected. A half mile up the western shore, near other mounds, I gathered numerous specimens of broken pottery of different moulds and colors. All the indications point to this place as being formerly a permanent rendezvous of these lost people.

Proceeding upon our very interesting and instructive voyage, other relics of the Mound-Builders were examined, notably those at Sandy lake, where copper spears and other useful and ornamental articles have been found many feet below the present natural surface of the earth. These are the waters so extensively used by the French and English, and later by the Americans, in portaging from the waters of the Great

Lakes to the waters of the Mississippi.[6]   It may well be considered as more than probable that the Mound-Builders were
the first to discover and utilize this great portage from the
basin of the St. Lawrence to the basin of the Mississippi,
and the Ojibways came after them in their encroachments,
which finally drove out the Sioux, who were the succeeding
race of men after the disappearance of the mound-building
people.

On the south shore of Sandy lake are visible the old landmarks of the trading post and station of a hundred years ago
which Lieut. Z. M. Pike so carefully described in his report of
the voyage of 1805-6 up the Mississippi, during the adminis-
tration of President Jefferson.[7] It is now an abandoned waste,
soon to be obliterated farther by the flood from the government reservoir dam about to be completed.   Two miles away
we found the site of the post and station occupied by the
Americans upon the acquirement of Louisiana from Napoleon
Bonaparte, with the acquiescence of the English government.
This old post and stockade was situated upon the east bank
of the Mississippi, and the well-preserved extremities of the
timbers used can be excavated from below the surface of the
earth, silent, inanimate reminders of the activities of the eighteenth century in maintaining traffic facilities with the tribes
in the then Far West.[8]

After having been joined by Dr. G. R. Metcalf and his son
at Winnibigoshish (most agreeable companions), our voyage
was continued down the river and brought to an end.

Soon afterward a return overland journey to the source of
the Mississippi was accomplished. On my return to Itasca
lake, I was firmly of the opinion that it had been discovered
by prehistoric man; yet years of casual examinations, from
time to time, since and including 1888, had failed to bring to

[6]This great portage was accomplished by passing up the St. Louis river to the
Dalles, thence by land to one of the several rivers, a few miles to the westward,
flowing into Sandy lake, which is scarcely a half mile east of the Mississippi
river, and connected therewith by a channel of unusual depth. The St. Louis
river is the most central and direct upper branch of the St. Lawrence river, the
main stream of a hydrographic system which includes the great fresh-water seas
of North America.
[7]Dr. Coues will, in his new "Pike," soon to be issued and published by
Francis P. Harper of New York, describe very fully the cause and results of the
Pike expedition up the Mississippi.
[8]Ojibways, Sioux, Mandans, Assiniboines, and other tribes and bands occupying the plains and territory from the Upper Mississippi west to the base of the
Rocky Mountains.

light any of the relics or landmarks of these lost people in that locality. These casual examinations had been made at the request of a distinguished geographical and historic writer, the late Mr. Alfred J. Hill, with no success whatever. It was now determined to commence a studied and careful exploration of the shores of Itasca lake for evidences of the existence there of man in the past ages. All the conveniences necessary for a two months' sojourn were provided, and the protection of the interests of the commonwealth against marauders at the state park gave the coveted opportunity to search thoroughly for some clue to proceed by and follow up. The success of this determination and the results which followed were surprising. The hydrographic and topographic surveys made on behalf of the State of Minnesota and its State Historical Society had been conducted under my personal supervision and direction, and I knew the locality better, probably, than any person of the present generation. At the time the final state park chart of 1892 was completed the words, "Earliest probable occupants, prehistoric," had been placed as a footnote in the legendary description, for it was at that time surmised that some day the opinion then entertained, which was the only basis for this legendary information, would prove to be well founded.

The night of the 26th day of October, 1894, the little animal, locally known as the pocket-gopher, which never sees the light of day except while throwing up in its peculiar way the surplus earth from its burrowings in little miniature mounds above the surface, made several of these well-known and peculiar mound markings a few feet above the surface of the water in Itasca lake on the east shore of the north arm, half way between McMullen's cabin, where I was encamped, and Patterson's old cabin, a quarter of a mile to the northwestward towards the outlet of the lake. On the morning of the 27th I discovered an unmistakable pottery remnant, which had been thrown up by this little pocket-gopher. This remnant of pottery bore several of the well-known markings of prehistoric man, peculiar to his residence in the Valley of the Mississippi. Thus the little mound-builder with his pouches, one on either side of the neck, extending from near the jaw down to near the shoulder, which we here designate by the very correct descriptive appellation of pocket-gopher, unconsciously brought to light the existence of the ancient mound-builder of more formidable portentousness, and who preceded this particular one by many centuries at the source of the Mississippi.

Now commenced a careful examination of the whole locality for further evidences, if such existed. First, a grooved stone hammer was found, then several additional pieces of pottery came to light in the stratum of the cultivated field belonging to Mr. McMullen, and on the 1st day of November Mr. F. J. Steinmetz came to my assistance and we prosecuted the search in earnest. A very old flint arrow-head came to light from the stratum near Patterson's old cabin; then two stone knives with well-defined and symmetrically chipped edges were unearthed in the immediate vicinity, and numerous pieces of broken pottery, of various unique and characteristic moulds, thrown up by the plough, hoe and spade, were added to our collection; then, not the least, by any means, a copper disk rewarded our patient search, soon after which followed a discovery of the unmistakable signs of a workshop, where were gathered the translucent and crypto-crystalline spalls struck from the prehistoric spear and arrow heads as they were made upon the shores of Itasca lake, and further over towards the outlet the white earth and decaying remnants of a shell-heap, long since covered by the mould and debris of ages, was definitely located on a point of ground above the surface of the lake. The ploughshare had thrown some of these decayed shells to the surface. After a systematic examination, it was concluded that a former village of Mound-Builders, nearly or quite one-half a mile in length, had been established and maintained in the Itascan region, in north latitude 47° 14' 15", longitude 95° 11' 41" west from Greenwich. From all indications it would appear that the cacique occupied the southeastern limit of the village, the better members, or head men, the center, and the lesser or poorer class the west end and flats there situated. This good guess, it is hoped, is very near the facts, for all of the better specimens and finer moulded articles appear at this supposed cacique's end of the village, the substantial mementoes came from the middle ground, and every piece or relic found at the westerly end was rough, poorly marked and of undoubtedly a skimp manner of making, indicating mediocre ability to finely mould. The chert and quartz and the copper unfold a remarkable and wonderful narrative of the geographical ability of these lost people. That they were geographers of no mean ability, courageous and mentally competent and able, can easily be surmised from these unmistakable evidences of their having penetrated to the heart of an unknown continent, without any subsistence, presumably, save only the results of their own ability to gain from a massive, unknown and dense wilderness. It was nearly three hundred years from the time the Mississippi river was known to exist by Europeans until Schoolcraft, in 1832, discovered and named Itasca lake. This lost village upon this spot was maintained at a time since which the sands of the earth and the mould of ages, with varying winds and storms and seasons, have accumulated over these deposited relics several inches, from no other than natural causes and at the summit of sloping ground. A forest of heavy timber has long since disappeared, leaving only the distinguishable evidences of where massive pines once stood. The copper, quartz and chert were undoubtedly obtained from the neighborhood of Lake Superior and other remote localities. Nothing whatever except the imperishable relics and the skeletons of this lost race of men remain, and they had only their hands and their wits by which to maintain themselves and their families in this solitude. I for one take the greatest interest in these remarkable people, who first penetrated to, and probably originally discovered, the source of the Mississippi. That they knew every hill and valley, lake and stream, at the Itasca basin, is shown by this extinct settlement of the dead, maintained previous to the origin of the North American Indian as found by European voyageurs.

Snow and ice put a stop to these explorations by the middle of November, but the collection of relics induces considerations and imaginations concerning these ancient people and their ability which can and will be augmented by a continuation of this fruitful and interesting search for more of these extant and imperishable evidences. Much remains to be unearthed. The whole course of the Mississippi river was occupied by these or similar tribes of men of ancient times.

The name by which they knew this great river, their language, religious ideas, marital habits, color, origin, much of their manners and taste, and the true appearance and construction of their lodges in this northern region, the mode of communication with other and distant villages, the habits of the chase, and all those personal characteristics necessarily peculiar to this race of men, must forever remain unknown, except in so far as we may be able to draw inferences, form opinions, and arrive at conclusions, after this whole western country shall have been searched for a more complete knowledge concerning these lost people, and then much must necessarily remain in the darkness of oblivion. Whence did they depart and what became of them? Who came after them and whence are these later people disappearing? The answer of this last question is nearer a solution than can possibly be claimed for the former.

V.

THE DAKOTAS AND THE OJIBWAYS.

M. Groseilliers and M. Radisson, two Frenchmen of energetic habits but apparently illiterate minds, about two hundred and thirty-four years ago, passing west from Lake Superior, came in contact with the Sioux or Dakotas, and as it is quite certain that these two first Europeans reached and crossed the Mississippi some thirty or forty miles above the present site of the city of St. Paul,[9] the gradual retirement of the Sioux before the aggressive Ojibway can be fairly traced from the happenings subsequent to that time. There is little doubt but that the two Frenchmen named, who at one time carried on their explorations under British auspices, were the first Europeans who came in contact with the Sioux tribes. They then lived in great numbers in the territory which afterward fell into the hands of their mortal enemies. While the full facts are not known, it is probable that the Sioux then occupied the entire waters of the Mississippi, from the region of the St. Croix to the source of the river, a distance of more than six hundred miles, with the adjacent country literally swarming with buffalo, elk, deer, bear and beaver, upon which

---

*Thirteen miles, by the channel of the Mississippi, below the Falls of St. Anthony of Padua, discovered and named by Hennepin.

they subsisted in comparative comfort. They do not know their own origin, and their legends scarcely indicate the facts of their migration to the source of the Mississippi. Whether they were the first to follow the Mound-Builders seems to remain a mystery. However, they came into possession of the country west from the extremity of Lake Superior, and remained there until they were, by force of arms, driven out by the Ojibways.

These later Indians considered themselves "spontaneous man" (An-ish-in-aub-ag). Their traditions, according to a learned writer of their own people, Hon. Wm. W. Warren, a mixed-blood, indicate that the meaning of the word is "Ojib," to pucker up, and "Ub-way," to roast. "To roast till puckered up." This seems to come from the manner in which they roasted their enemies until they puckered up. Another interpretation is the manner in which they pucker up their moccasins in seams below the instep. Mr. Warren intimates that they may have descended from one of the lost tribes of Israel, suggesting Hebrew extraction, and their first known residence was not far from the mouth of the St. Lawrence, on the coast of the Atlantic. Their migration westward necessarily covered centuries, for, tarrying a long time in the neighborhood of the outlet of Lake Superior, they afterward resided upon the Island of La Pointe for over a hundred years, near the Bay of Sha-ga-waum-ik-oug (Chaquamegon Bay, Lake Superior). Here their extensive rendezvous, located upon an island to escape the onslaught of their warring enemies, at a time when firearms were unknown to them, was broken up by cannibalism among themselves, and the numerous clans of the tribe scattered in different directions. Coming into the use of firearms, they pressed their warfare against the Sioux until they came into possession by force of arms of the entire upper waters of the Mississippi north of the mouth of Watab river, immediately above Sauk Rapids, Minn., and a large area in the valley of the Red river of the North. This war of unknown duration almost transformed certain habits of the Sioux, for they departed permanently from the timbered localities near Mille Lacs, Sandy and Leech lakes, and soon became a people of the treeless plains, reaching from the valley of the Minnesota river to and across the basin of the Missouri in South and North

Dakota, using ponies for transportation purposes,[10] while the Ojibways made use of the bark canoe and pack-strap until very recent years. This remarkable history of the Sioux and Ojibways, if given in detail, would fill volumes. The last war party between these contending tribes of which I remember was of the Sioux in 1860, from the valley of the Minnesota river to Crow Wing river. For considerations of vital importance, a full description of which is dispensable in an article of this kind, these Indians of both tribes are disappearing

[10] While preparing for a desultory march, the Sioux Indians, who occupied the treeless plains for so long a period, fastened two long lodge poles to either side of a mustang pony with long busby tail, one end of the pole resting against the shoulder of the animal and the other on the ground, from six to eight feet in the rear of the pony. These poles were fastened by broad straps made from buffalo or elk skin used as breast straps, and crosswise at the rear end of the lodge poles were fastened two shorter poles, forming a square frame. To this frame would be fastened the skin of a buffalo, usually in rawhide form, which completed a unique, ingenious means of transportation, impossible to upset in the rugged passages of the wild west. Loading dried meat, pemican, cooking vessels, blankets, papooses, etc., on this square frame of poles and rawhide fastened to an unruly pony, without bridle, driver or harness, the whole was turned loose as being ready for the march. The women of the band, invariably designated by the euphonious appellation of "squaw," attended these ponies on foot. As a rule they were poorly dressed, wearing moccasins without stockings, short garments, leggins of cloth or leather made from skins, bareheaded, with long black braids of coarse hair reaching down the back, usually a calico waist and short skirt, and during inclement weather a square blanket for a wrap and hood, which, when thrown over the head and wrapped around the arms and body, would leave only the face and feet protruding. These women were the laborers and servants of their husbands and masters. An Indian seldom pitches camp, loads the pony, cuts fuel or carries water, when the squaw, wife and mother accompanies the moving band. With a painted feather in his hair for each scalp taken, the best pony of the herd, with skin saddle, decorated with colored beads, stirrups of rawhide or thongs, and a bridle made from elk skin; bowie knife, scabbard, gun, pipe and kinnikinic—the pipe of stone, the kinnikinic gathered from the bark of the red willow, held in a beaded tobacco pouch—with punk and the accompanying flint and steel with which to strike fire (the steel in the right hand and the flint and punk held tightly with the forefinger and thumb of the left hand, when in use); a calico shirt: no hat or cap; leggins fastened from above the knee to the belt, which carries the knife, tomahawk, etc.; a breech-clout, over which loosely hangs the calico shirt; beaded moccasins and a blanket (usually white or green in color), and a miscellaneous outfit of trappings, ammunition for the chase, painted face, vermilion on the hair where it is parted in the middle, gaudy ornaments of a cheap variety, braided hair, no beard, a skin that is nearer black than it is red; thus mounted he marches a prince of the plains, ready for war, the dance, the hunt, the leisurely smoke, his daily decorations, but never the degrading, despicable labor of the camp. The striking appearance of a band of Sioux Indians, marching in the manner described, can only be adequately understood and appreciated by those who have witnessed these actual scenes in the years gone by; for now these Dakotas are no longer roaming nomads, but are housed on reservations in the Missouri valley and fed by the government.

The Ojibways are but little different, save only that they were a people of the woods, using birch-bark canoes, traversing the numerous water courses, portaging by the use of the pack-strap, made from the thick heavy skin from the leg of the moose, by which means, his bark canoe, bottom up, balanced over the top of his head, with the bow of the canoe sufficiently elevated to permit him to look forward under it, was carried from lake to lake and from river to river. The squaws, by the same kind of pack-strap, carried upon their backs, balanced from the top of the head, the camping outfit, cooking utensils and paraphernalia, with the papoose strapped on top of the whole, face to the rear, in a frame to which it is tightly bound. The canoe carried by the braves weighed about fifty pounds, while the load carried by the women often weighed three hundred pounds. Their portages were never very long, but in the winter season their marches upon snow-shoes were often a hundred miles or more. Of recent years these Indians are housed upon reservations, and have teams and wagons furnished by the government. The former appearance of the Ojibways on their marches was no less striking, though different from the Sioux in the manner described in this brief note.

A CANOE AND OJIBWAYS.

VOYAGING ACROSS LEECH LAKE.

OJIBWAY GRAVES, AT LEECH LAKE.

from the face of the earth. Warfare, immoral habits, small-pox, the inebriate's weakness, consumption and miscellaneous degenerating influences have depleted their ranks to an extent which makes the final result concisely rapid and silently sure. There are reasons why they should still continue to regard the white man as a mortal enemy.

There are two known destinies: the disappearance of the Mound-Builders and the disappearing Indian tribes of the upper basin of the Mississippi, for the second and third race of man[11] known to occupy and inhabit the upper basin of the Great River are following in the footprints of their one known predecessor in an assured disappearance, unless all signs fail; and the pale-faces—"Not Frenchmen, nor English, but white Indians"[12]—are now the active, ambitious, energetic occupants of the entire basin of the greatest river system of the world, with the simple exception of isolated reservations,[13] and most of these will soon be possessed by a hardy pioneer people.[14]

[11]The Sioux and the Ojibway tribes, distinctly separate, but probably of nearly or quite the same very remote origin.

[12]Lieut. Z. M. Pike and his soldiers, in 1806, were designated by the Ojibways "White Indians," because they were neither Frenchmen nor Englishmen, as was usual in those days, but of that American nation of men to whose existence the attention of the Ojibways had not been directed prior to Pike's visit to them.

[13]All Indian reservations in northern Minnesota are Ojibway, and all in the Dakota states are Sioux, excepting the Turtle Mountain reserve. Several of the Ojibway reservations have been transferred to the public domain by congressional enactment, and the Indians ordered removed to the White Earth reservation as a permanent place of abode. The former policy of the government of the United States in dealing with these Indians was by formal treaty, but recently a change has been inaugurated, suggested, I think, by the late Gen. B. F. Butler, whereby the authorities of the United States no longer consider the Indians a proper people to treat with.

[14]Citizens of the United States, among whom are a considerable number of Scandinavian, Danish, Finnish, German, Polish, and other people, emigrants from the shores of Europe, a well-to-do class as a whole, honest, industrious, and capable of exercising the rights of freemen under a government in the temperate zone, the brightest and ablest men of which were born and reared in log cabins. By a reference to any chart showing correct geographical positions, comparable with statistical results, it will be noticed that the central intensity of the north temperate zone encircles the earth immediately in the neighborhood of the Great Lakes in the western hemisphere, where ample elevation above the sea level, pure air and water, wholesome food, and the consequent activity and development of the mind and body, produce a race of men as yet unsurpassed. These climatic and hygienic influences soon transform the languid emigrant into an energetic citizen. To this same influence I attribute the success of prehistoric man in this same locality (the basin of the Mississippi), in energetically pushing forward for a more extended geographical knowledge, until he built and maintained a town at the very source of the river, in the exercise of the laws of existence, showing a knowledge of latitude and departure, exceeded only by the use of scientific instruments of a more modern advantage. There is little or no doubt in my mind that the remarkable progress of the American people and the wonderful strides made by them toward a revolution in scientific research and invention come from this intensity of the temperate zone in its capacity to enlarge, expand, enhance and characterize brain formation at geographical positions where the greatest power of the sun's rays is intensified by other and consequent subsidiary influences; for certain it is that the arctic and the tropic zones have produced no such transformation in the world's history as has the federal republic in this temperate hemisphere in but little more than a century of time. Then it is reasonable to presume that the Mound-Builders, in their physical and mental capacity, were intensified by the same climatic influences which gave them the energy to discover the source of the Mississippi, of which fact we are now made aware, as a result of this recent and very interesting exploration.

VI.

RESULTS OF THE EXPEDITION OF 1895.

The close of the year 1894 witnessed the new discoveries related in the preceding pages of this communication. The relics and remains of the Mound-Builders at Winnibigoshish and at Itasca lakes, apparently deposited at about the same period of time, left no doubt of a more extended occupancy between the two points named, adjacent to the numerous lakes and streams which extend throughout the upper watershed of the Mississippi.

It was toward a solution of this latter problem that I gave my attention at the beginning of the present year. In the month of February last the late Mr. Alfred J. Hill again became my associate, preparatory to a more extended and systematic exploration of the different positions between and adjacent to the localities named, a distance by the channel of the river of a little more than one hundred and thirteen miles.

All preparations necessary for this third voyage of discovery were made by myself, and Mr. Hill delegated his portion of the work to the able hands of Prof. T. H. Lewis, who became an equal party to these new explorations at our joint request, Mr. Hill himself having determined that he could not personally accompany me. The movements of the party during the time occupied are here given in a narrative description of the explorations and discoveries made. There seems to be no necessity for any distinction between the particular facts discovered by Professor Lewis or myself, for upon every hand we jointly or separately brought to light a most remarkable and deeply interesting list of discoveries connected with and bearing upon the occupation of the entire upper basin of the Mississippi by prehistoric man.

The expedition proceeded to Park Rapids, Minn., with a very complete supply of surveying instruments, maps, charts, camera, government plats, one boat and lighter, and all necessary provisions and apparel for a two month's voyage in the northern wilderness. It was on the 27th day of April, 1895, that a further and rather cursory examination was made of the Dickenson mounds, situated upon the south side of section 14, township 140, range 35, one mile north of Park Rapids, Minn.

We counted in this group twenty-two various mounds and embankments, of different sizes and heights. These ancient works are situated southwestwardly from the outlet of Fishhook lake, and less than one mile distant therefrom. There are slight indications of a village site of Mound-Builders on the south side of the lake near these mounds, and at the farmhouse of Mr. Phipps are two other mounds, near the west end of the lake. Chipped spear-heads and arrow-points of stone were exhibited by Mr. Phipps, a collection gathered in his field, which discloses to a certainty that the mounds near Fishhook lake and the Dickenson mounds and earthworks were constructed by the prehistoric mound-building race. No detailed survey of these earthworks has been made. On the evening of the 28th of April the members of this expedition established an encampment at McMullen's, on the north end of Itasca lake, and until the 5th of May explored, from day to day, the entire surroundings of the prehistoric village site discovered by me the previous October. Numerous arrow-points of stone, pottery shards, spalls and chipped stone implements were found on both sides of the Mississippi and along the east shore of the north arm of Itasca lake, indicating that the former ancient occupancy was more extensive and of greater age than was at first apparent. Our next discovery was the site of an old trading station of former years, date unknown, situated upon Schoolcraft Island. This old station, unmentioned by any of the earliest explorers, was probably a trading post of the French in early times, and I have referred the matter for some further inquiry to Professor Levasseur of the Department of Public Instruction for France.

A group of ten burial mounds was discovered upon the fractional east half of the south-west quarter of section 35, township 144, range 36, which I have properly named in honor of my discovering companion. A more detailed description of these mounds, mapped in plate VI., follows herewith:

THE LEWIS MOUNDS.

1. Diameter eighteen feet, height one foot.
2. Length eighty-three feet, width sixteen feet at the east end, twenty-one feet at the west end, height two and one-half feet.
3. An elliptical mound, length thirty-eight feet, width twenty-four feet, height three feet.
4. Diameter seventeen feet, height one and one-half feet.

5. Length forty-three feet, width sixteen feet at the west end, twenty-four feet at the east end, height two feet, about the shape of an egg cut in two lengthwise, and the half shell turned down.

6. Diameter twenty-six feet, height three feet.

7. Diameter twenty-two feet, height three feet.

8. An elliptical mound, length twenty-eight feet, height two and one-half feet.

9. Diameter sixteen feet, height two and one-half feet.

10. An embankment, forty-four feet in length, eighteen feet in width and two and one-half feet in height.

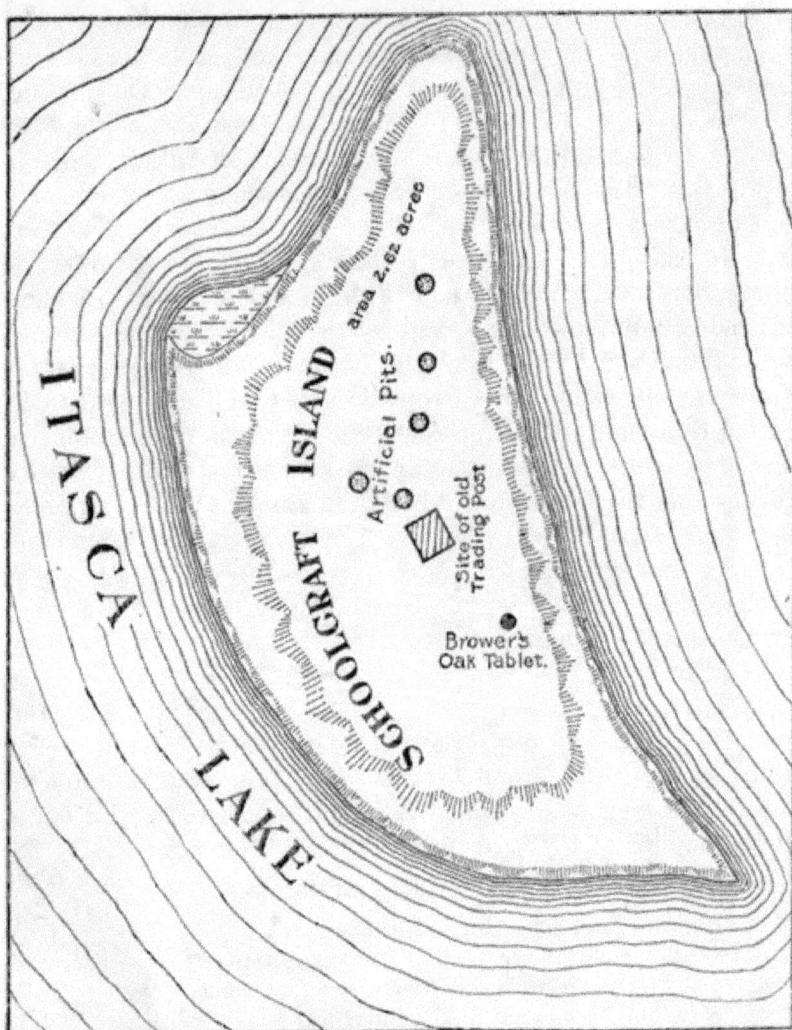

FIG. 1. SKETCH MAP OF SCHOOLCRAFT ISLAND, ITASCA LAKE.

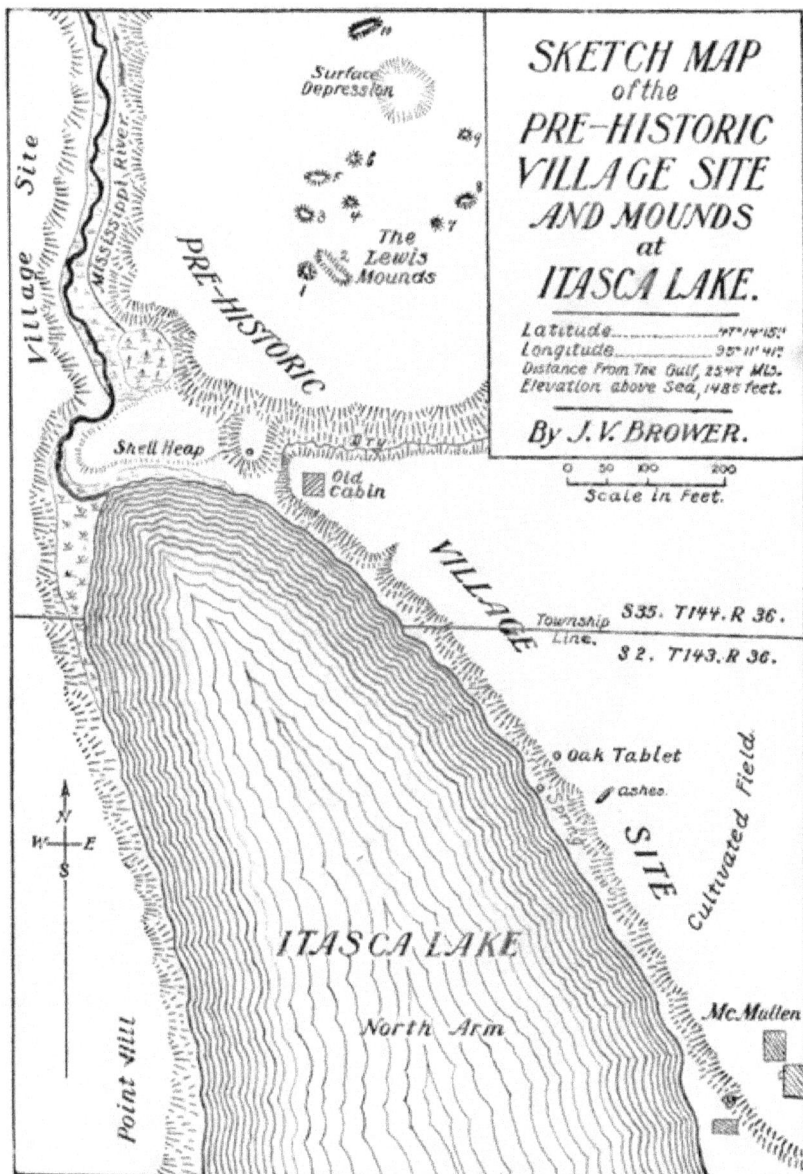

SKETCH MAP
of the
PRE-HISTORIC
VILLAGE SITE
AND MOUNDS
at
ITASCA LAKE.

Latitude............47°14'15"
Longitude............95°11'41"
Distance From The Gulf, 2547 Mls.
Elevation above Sea, 1485 feet.

By J. V. BROWER.

0   50   100   200
Scale in Feet.

Village Site

Village Site

Mississippi River

PRE-HISTORIC

Surface
Depression

The
Lewis
Mounds

Shell Heap

Old
cabin

VILLAGE

Township
Line

S 35. T 144. R 36.
S 2. T 143. R 36.

Oak Tablet
ashes
Spring

SITE

Cultivated Field

N
W   E
S

ITASCA LAKE

North Arm

Point Hill

McMullen

MAP OF THE LEWIS MOUNDS AND PREHISTORIC VILLAGE SITE AT
THE NORTH END OF ITASCA LAKE.

With the assistance of Messrs. Wegmann and Sauer, whom we engaged for the occasion, several of this interesting group of mounds were excavated, with the following results:

EXCAVATIONS.

Mound No. 1 was composed of sandy loam. The remains of one or two interments in this mound were fragmentary and useless for scientific comparison.

Mound No. 2 was not excavated.

Mound No. 3. Composed principally of black sandy loam. At the west side of the center the loam of the original surface had been removed. Resting upon the natural gravel below this excavated loam was a quantity of calcined human bones. Five skulls were recognizable and the fragments of probably as many more were intermixed in this heap of charred remnants. At the north edge of the calcined remains was a well-preserved skull. Just above this calcined mass of human remains and almost resting upon it were six skulls and various bones, more or less decomposed and broken. Still above these last described remains and near the upper surface of the mound appeared the remains of an intrusive burial of doubtful identity; but since a well-defined covering of birch bark appeared, this latter interment was undoubtedly by Sioux or Ojibway Indians, probably the latter. The remains of this last interment were very much blackened and decomposed, while on the other hand the skulls lower down in this place of burial were natural in color; a comparison in the mode of burial which presents a wide difference. At the east end of the excavation there had been buried the remains of seven persons, but throughout the extent of the excavation there was wanting any evidence of regularity in the mode of burial. In different sections of the mound two small beds of gravelly sand and two of charcoal and ashes were noticed, but no certainly defined existence of fire at the time of burial could be traced. A portion of the bones were calcined.

Mound No. 4 was composed of black sandy loam, and contained the disappearing remains of but one person near the bottom of the mound.

Mound No. 5. Composed of a light sandy loam. Near the east end a small pit, five feet in diameter, had been excavated below the original surface about one and one-half feet. From this artificial pit there were taken three skulls and a few bones, very much decayed and broken. At the east end appeared a quantity of debris, consisting in part of broken bones, pottery shards, charcoal and ashes, but the bones were not of human origin.

Mound No. 6 was composed of sandy loam, and contained, apparently, the fragments of two decayed skeletons.

Mound No. 7. Composed of sandy loam. Only one pottery shard was found in this mound.

Mound No. 8. Composed of sand and sandy loam. Two small ash heaps and a few fragments of human remains only were found in this outlying place of burial.

Mound No. 9. Composed of sandy loam. Near the surface were two intrusive burials, male and female, and the same considerations apply to these which appear concerning the upper burials in Mound No. 3. I am of the opinion, however, that these are the remains of Ojibway Indians, buried near the surface, in the flesh, and not, therefore, prepared for a continuous preservation as were the calcined remains of the dead Mound-Builders, interred so long ago in the mound referred to. The other remains in this mound had long since crumbled to dust.

Mound No. 10. Composed in part of a sandy clay and sandy loam. Near the center of this mound were two skulls and parts of three skeletons. Beyond a trench, about twenty-eight feet in length, run through the upper part of this place of burial, nothing of interest appeared. The interments were original.

Commencing at the site of the central portion of the Lewis group, and extending to the Mississippi river on the west and to Itasca lake on the south, there appeared numerous stone spalls and pottery shards, indicating beyond doubt a more defined outline of the village site maintained there during the centuries long since passed.

At Point Hill, Itasca lake, named by Dr. Coues in honor of my late distinguished associate, there was discovered one mound twenty-four feet in diameter and two feet in height, which contained fragments of bone and mussel shell. At the summit of the south end of Point Hill, a remarkable bone heap was excavated, about twenty feet above the surface of

EIGHTEEN SPECIMENS OF POTTERY SHARDS FROM ITASCA LAKE.

the water in the lake. I noticed bones of the moose, bear, deer, wolf, beaver, and fox; and intermixed therewith were fragments of pottery, stone spalls, hearthstones, and triangular arrow-points, indicating the former existence of a small village of Mound-Builders, probably at about the same time that the extensive village on the north end of Itasca lake was maintained. Taking advantage of our sojourn at McMullen's, a large collection of relics was made, illustrated partly in plates VII. and VIII.

On the eighth day after our arrival at Itasca lake, we departed northward by team and boat and camped for the night of the 6th at the Shanafelt Bluffs at section 30, township 145, range 35. On the morning of the 7th our course of departure was down the Mississippi in our comfortable klinker. Upon arriving at the mouth of Chemaun creek, an examination of the surrounding country was made and the existence of pottery shards at lot 12, section 19, township 145, range 35, was noted. There was also found on the east side of the creek a very old pipe of red pipestone, the identity of which is uncertain. We camped for the night at my former camp (Trouble), at lot 9, section 5, township 145, range 35. Proceeding on our voyage we camped for the night of May 8th about two miles above the mouth of the Piniddiwin river. The following morning a very cursory examination of the hills bordering upon Manomin lake, through which the Piniddiwin takes its course, revealed meager signs only of the migratory pathway of the Mound-Builders, but enough to satisfy us that they had formerly occupied the premises.

Continuing down the river, which meanders through the extensive meadows to the eastward, we passed to the first plateau below the mouth of Cow Horn creek and landed for a noonday lunch at the fractional northwest quarter of section 28, township 146, range 34, Beltrami county. Here we discovered the evidences of a former large village site of the Mound-Builders on both sides of the Mississippi. Numerous pottery shards, spalls, and one stone scraper were found. A search for the mounds of the locality being unsuccessful, we proceeded on our way and camped for the night on the north bank of the river, at lot 1, section 27, township 146, range 34. Proceeding upon our voyage, we discovered numerous evidences at the extensive sand-bank upon section 13, and afterward it was learned that there was a large mound in the same neighborhood.

The
Cow Horn
village site.

NW¼
Sec 23. T 146. R 34.
Minn.

FIG. 2.

At Carr's field, at the mouth of Naiwa river, where the same unites with the Mississippi, at the fractional south-east quarter of section 20, township 146, range 33, numerous evidences of a large prehistoric village site were selected from the upturned earth in the cultivated field there situated. We voyaged up Naiwa river through the first lake and camped on the east shore of the second handsome body of water, first north from the picturesque Plantagenet, which is the Resting lake of Allen's map. The confusion of names which Dr. Coues so strikingly illustrates for his new "Pike," in a valuable historico-geographical chart, an advance copy of which is open before me at this writing, admonishes me not to undertake the task of unraveling the classified nomenclature of this locality during a consideration of this present subject.

SPECIMENS OF RELICS COLLECTED AT ITASCA, BEMIDJI, TASCODIAC, COUES, CASS, AND LEECH LAKES.

On May 11th, with varying winds, we reached an encampment at Bemidji lake, on the east bank of the Mississippi, at its entrance into this magnificent body of water. Our encampment was the site of the encampment of prehistoric man,

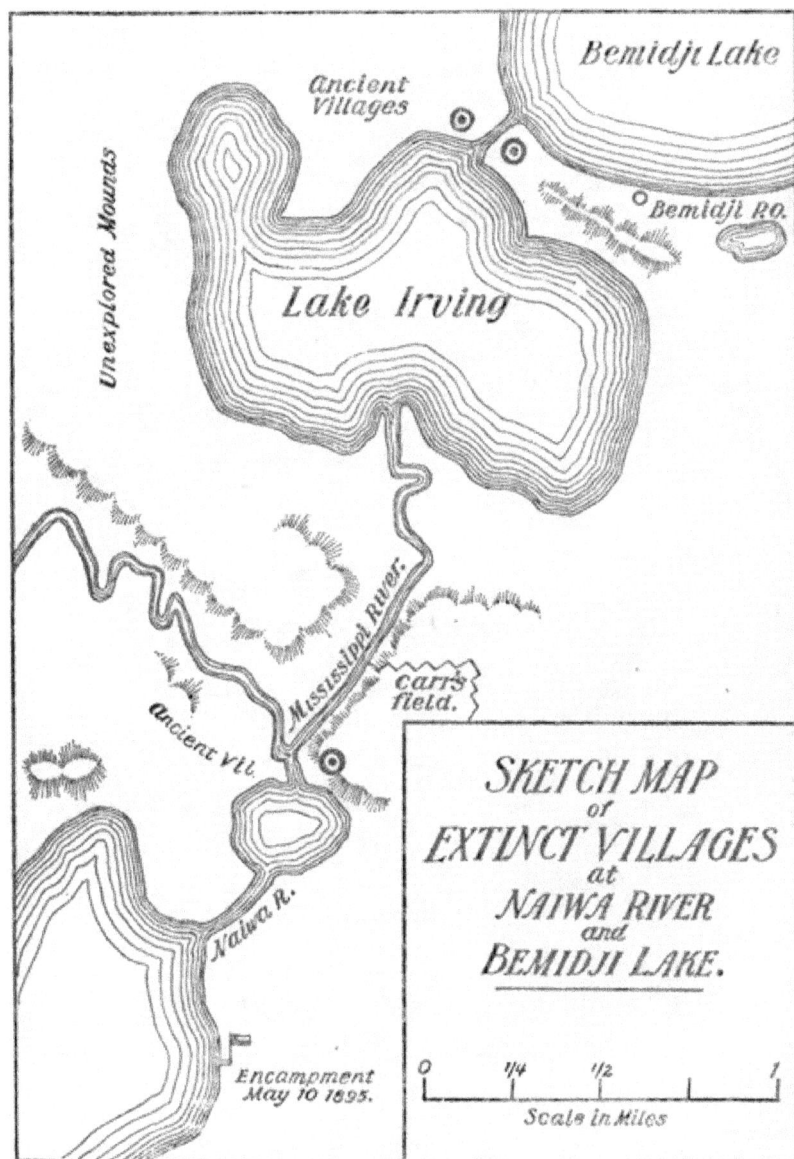

FIG. 3.

for on every hand and on either side of the river we gathered promiscuously the relics and remnants of the mound-building race of men, including a perforated cowry shell (Cypraea annulus). We learned that the mounds of this locality were

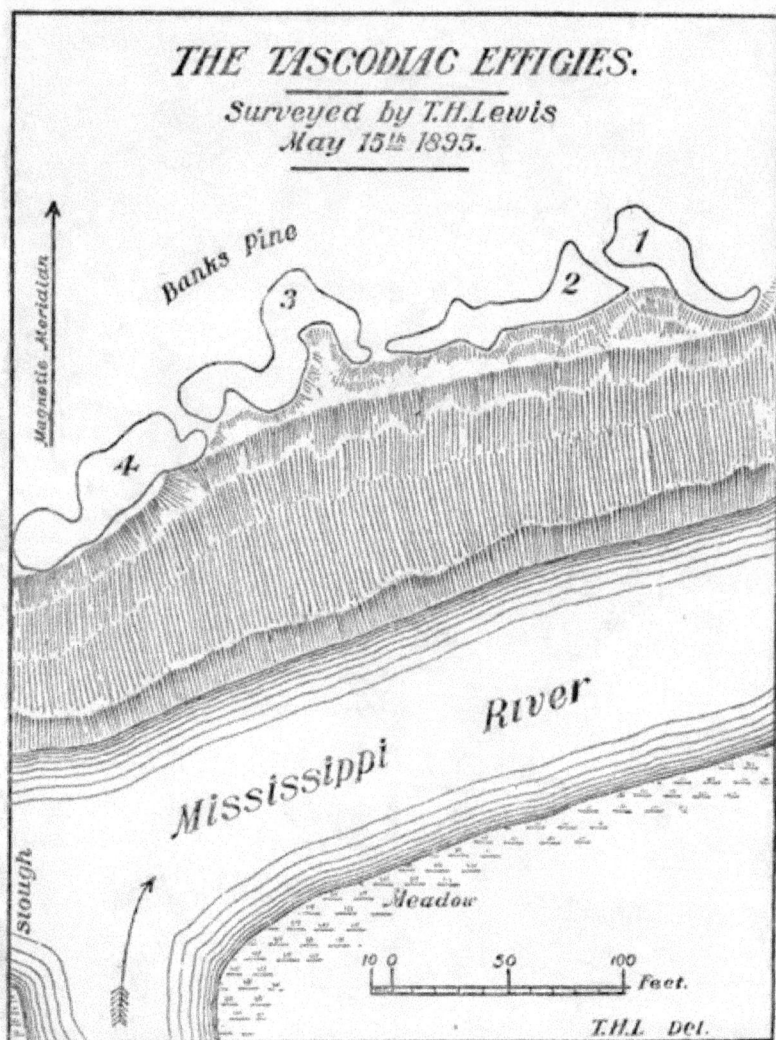

THE TASCODIAC EFFIGIES.
Surveyed by T.H.Lewis
May 15th 1895.

FIG. 4.

NOTE.—Immediately in the rear of Mounds 3 and 4 is a very considerable depression, several feet in depth, from which, it is possible, the earth was excavated for the construction of these interesting Tascodiac effigies.

situated at an eminence west of Lake Irving. I explored the unique geological ridge between lakes Irving and Bemidji, and entertain some reasons for believing that modified mounds are to be found on this ridge, which has been variously occupied by the Ojibways for a hundred years or more. At the outlet of Bemidji lake, on both sides of the Mississippi, relics of the Mound-Builders were picked up, and a large mound was discovered immediately at the outlet on the north bank of the river and east shore of the lake, near the base of a very old oak tree. This mound was partially excavated and found to be of black sandy loam, containing the remains of original interments, only one of which was removed, in a fair state of preservation.

In passing down the river, the most northerly course of the Mississippi at my camp of the year previous, known as Camp Boutwell, was reached and passed, and the fair stage of water in the river gave us great pleasure in voyaging over and down the numerous rapids extending from the Bemidji outlet to the locality of the Tascodiac. Here I take issue with Dr. Coues, who in "The Annals of Iowa" for April last, deplores the low water and impassable rapids of this portion of his voyage of 1894. That was a season of drouth, and now this particular portion of the Mississippi is the most romantic and picturesque of the entire upper basin of the river, easy to navigate and interesting to explore.

The night of May 13th found us encamped at a limited plateau on the west bank of the river, opposite a small grassy island, about three miles above Tascodiac lake. Pottery shards were found at our landing place. Proceeding on the morning of the 14th, we soon went into camp, on account of rain, at the edge of the Tascodiac meadows and within sight of the lake. Having discovered four extensive effigy mounds, at the summit of the bluff on the north bank of the river, the 15th of May was the time allotted to survey and excavate this curious group. Dr. Young, Dr. Shanafelt and myself had explored this locality on Monday, August 13, 1894, ascending to the summit of each of these mounds. They seem to have been constructed for some unknown purpose, out of pure sand, and the group contains over one hundred tons of earth. An excavation of the most southerly mound of the group, to the original surface, brought nothing to light bearing upon the

question of purpose in the construction of these old earthy
effigies.

Subsequently the village sites and mounds of prehistoric
Tascodiac man were located on both sides of the Mississippi

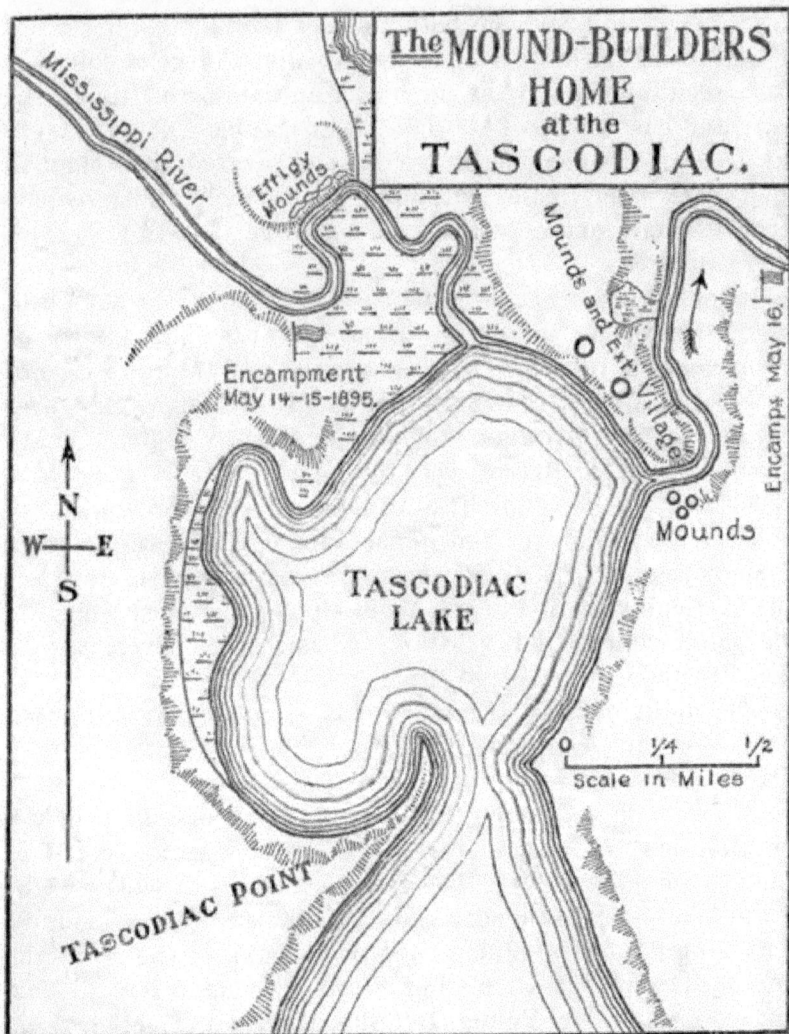

FIG. 5.

at the outlet of the lake, about a mile distant from and in full
view of the Tascodiac effigies. Toward these village sites and
mounds we extended a particular investigation. Stone and

THE MISSISSIPPI, ABOVE THE MOUTH OF THE NAIWA RIVER.

THE MOST NORTHERN ISLAND IN THE MISSISSIPPI, BETWEEN BEMIDJI
AND TASCODIAC LAKES

pottery remnants are promiscuously scattered about, along the sandy beach of the lake, upon lots 6 and 9, section 25, township 146, range 32, at the western boundary line of the Ojibway reservation. Two large burial mounds are central at the principal village site, on the point of land nearly encompassed by the lake and river, and on the south side several small low mounds appear at the summit of a hillock near an old trail leading from Leech to Red lake. The most southerly mound on the north side of the river was excavated, and disclosed a very interesting state of facts. This mound is forty feet in diameter, three and one-half feet in height, with an approach about two feet in height and thirty-six feet in length extending northwestwardly from the base of the mound. We here exhumed the skulls of twenty persons and portions of twelve others, which, with three additional ones noted at the side of the excavation, made in all thirty-five within a space dug down through the center of the mound scarcely seven feet in diameter. Other portions of these skeletons appeared in such a promiscuous manner, intermixed in such different and irregular order, that it leaves the cause and manner of this wholesale burial in doubt. Several large boulders were taken from this excavation, placed there by design for some purpose, usually above one or more skulls. That there were probably upwards of one hundred remains laid to rest in this particular mound seems possible, and the manner of burial with accompanying pottery shards and other prehistoric evidences leaves no room for conjecture as to the identity of the contents of these mounds. They were the builders of the Tascodiac effigies. There are many reasons for determining that they were. Some two hundred and fifty yards northwesterly appears another mound, nearly the same size and circular in form. The use of the steel probe indicated that this mound, like the first, is filled with human remains, but for want of time it was not excavated. Between, around and outside of these mounds were the remnants and debris of a former extensive village. Burned stones, chert, quartz, hornstone and jasper spalls and a few chipped implements were found, and pottery shards variously composed of stone and clay, sand and clay, and shell and clay, lie scattered along the sandy beach of the lake, which is modified somewhat, like the river bank above, by the action of water.

After a night's rest at our encampment on the south side of the river, a mile below Tascodiac lake, we proceeded down the stream to the Elliott Coues and Cass lake locality. Ojibway villages and settlements are variously scattered along

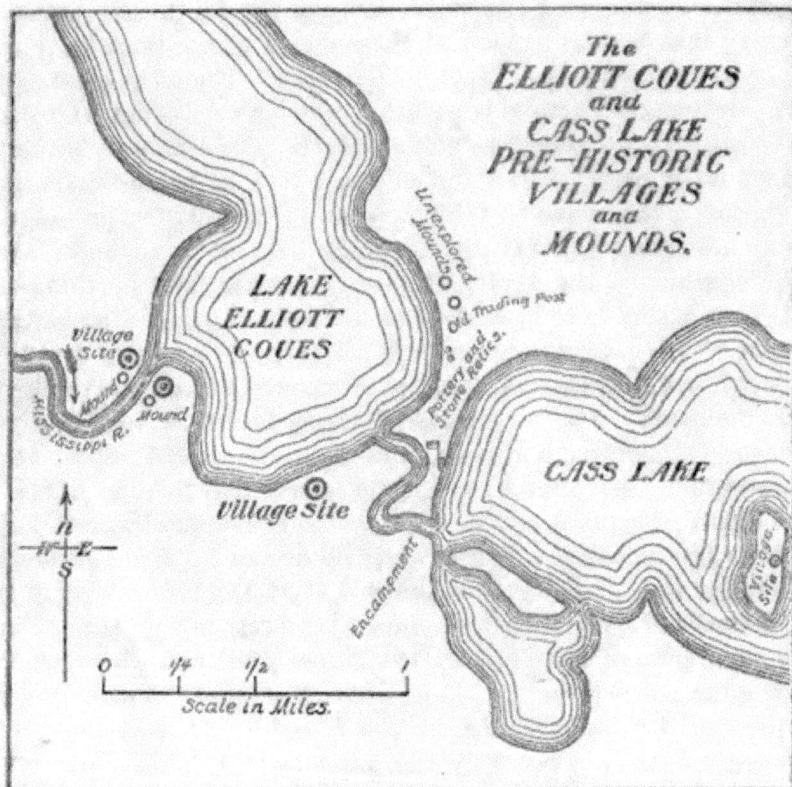

The
ELLIOTT COUES
and
CASS LAKE
PRE-HISTORIC
VILLAGES
and
MOUNDS.

LAKE ELLIOTT COUES

CASS LAKE

Unexplored Mounds

Old Trading Post

Pottery and Stone relics.

Village Site

Mound

Mound

Mississippi R.

Village Site

Encampment

Village Site

0   1/4   1/2   1
Scale in Miles.

FIG. 6.

NOTE.—Indications point with an unerring certainty to a probability, that the mound-building population occupied many localities at and in the neighborhood of Cass lake and the rivers and portages leading to and from it, which would require a considerable length of time to explore and survey. An important village site was discovered at the southern extremity of Pike bay, seven miles south of the islands shown in the above sketch map. Leading south from the Pike bay village site is a prehistoric trail or portage, along which were collected several relics. This trail or portage leads to the north shore of Leech lake, at one of the great central village sites of the ancient occupancy at Mound point.

the route from the entrance of the Mississippi into Lake Elliott Coues, along the north shore of Cass and Winnibigoshish lakes, to the mouth of the Cut Foot Sioux, something more than thirty miles. At the very inception of our route through and past these villages and settlements unmistakable evidences of the former mound-building population maintaining a permanent occupancy were observable on every hand; and the little cultivated fields established by the Indians proved what we had suspected, for everywhere were found the pottery shards, stone spalls and chipped implements of the prehistoric age. The Ojibway cemetery at the first Indian village reached, on the left bank of the river, situated upon lot 1, section 30, township 146, range 31, includes intrusive burials in one or more mounds; and on the opposite bank of the river, a little farther up stream, is a single mound about fifty feet in length and three feet high. There are also indications of artificial earthworks at the summit of a sharp declivity on the opposite side of the lake, north of our place of encampment, which we effected at the west extremity of Cass lake. At and immediately north from this encampment are the indications of the occupancy by man in succeeding ages for at least a thousand years, and possibly a much longer time. Ample evidences of a mound-building population were noticed; the Sioux Indians resided here; and a very old trading post was maintained, the fallen stone fireplaces only remaining at excavations which mark the spot upon lots 2 and 3, section 29, township 146, range 31. The ruins of an old mission are situated on the north shore opposite the island of Ozawindib, named after Schoolcraft's guide. The Ojibway population know no date connected with the coming of their forefathers, when the Sioux retired from this most central location of the upper basin, unable to withstand the onslaughts of their advancing enemies. The islands in Cass lake also plainly reveal a former occupancy by a prehistoric race, for there likewise can be found in abundance the same imperishable relics of pottery and chipped stone which exist at the other points along our route.

Owing to stormy weather we concluded to change our plans somewhat, and turned toward the portage from Pike bay to Leech lake. At the southern extremity of Pike bay, where we were encamped for a day, an extensive prehistoric village existed along the plateau there situated, and the stone spalls and pottery shards collected along the old trail from this point to the north shore of Leech lake yielded an abundant and interesting budget of information concerning the discovery and use of these old portages long centuries before Lieutenant Pike or the Ojibway Indians traversed the locality situated between the two points named.

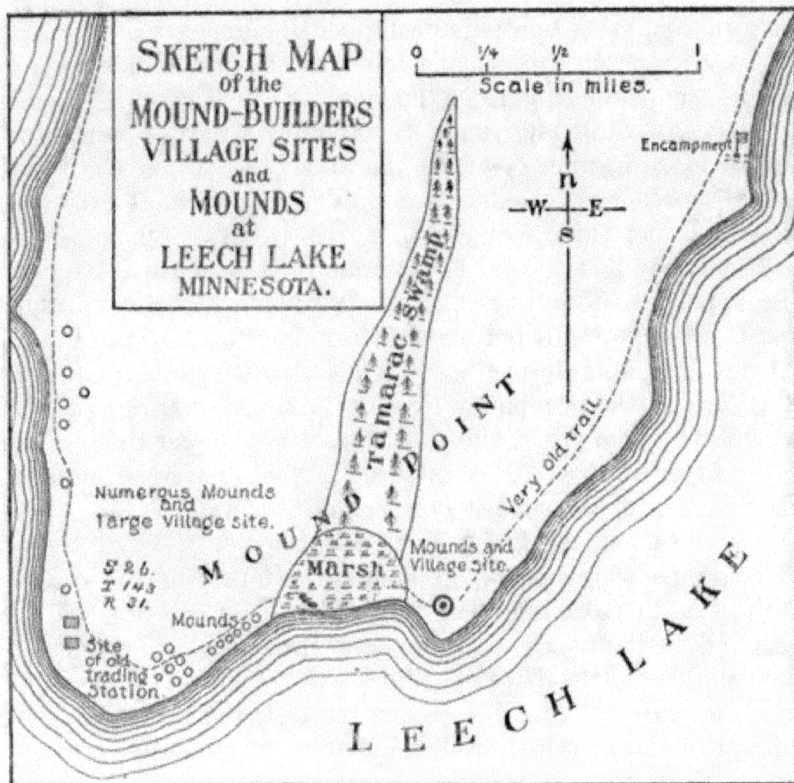

FIG. 7.

NOTE.—There is probably about one hundred miles of shore line at Leech lake, of which extension less than five miles was explored, creating the reasonable supposition, that, adopting Mound point as a criterion, there are nearly or quite two hundred mounds at and near this lake and the rivers flowing into it, which would require more than a month's time to properly explore and survey. The principal streams flowing into Leech lake are Little Boy, Shingobi, Ka-be-ko-na, and Bukesagidowag (or Steamboat) rivers, and numerous smaller lakes and streams are found in all directions. After the retirement of the Sioux Indians this locality was selected by the Ojibways, as one of several permanent places of abode, where they still reside, subsisting principally upon game and fish and the annuities paid by the United States under treaty stipulations, or according to congressional enactments.

Our investigations at Leech lake revealed the former existence of a great central village of Mound-Builders, situated for miles along the north shore, upon that broad point of land immediately east of the most northwesterly arm of the lake.

It may be safe to determine that the most central portion of this ancient town was situated upon sections 23 and 26, township 143, range 31. My accommodating Indian guide led the way to an innumerable line of burial mounds of different forms and sizes, some in groups, others scattered about, and some variously modified by intrusive burials. At one point I noticed modifications which suggested the possibility that they had been used as rice-pits by the native population. A handsome collection of relics at this locality rewarded our search. At one point the former existence of an old trading station was noticed, the old stone fireplaces marking the spot. Finding no name for this broad and extensive point of land heretofore applied, I have called it Mound point. The Ottertail point of Dr. Coues' chart of 1895 should be east of Goose island, where we also found ample evidences of the existence of a former mound-building population.

An unfortunate misunderstanding having occurred between my companion and some native Indians, who object to surveying operations upon their reservation, we removed from this interesting locality to the southwest arm of the lake and thence to the mouth of the Shingobi river, where we were storm-bound; and this change of base was made without a coveted exploration of the Ka-be-ko-na lake and river, at the west side of Leech lake, where a prehistoric occupancy existed, probably quite as extensive as that of Mound point. This information we gained from our Indian neighbors.

The ascent of Shingobi river was to me an important event, for, reaching the portage to the Crow Wing lakes and river, which crosses the Itasca moraine, I followed the deep trail to the east shore of a small lake, across which, the next day, we discovered a well-defined prehistoric village site and mounds, away from which led, toward the westward, the same trail to the Crow Wing river. This old village site and mounds are located about halfway between the Shingobi and Crow Wing rivers, but upon what particular section we did not determine. It will be remembered that Mr. Schoolcraft and Lieutenant Allen passed over this identical portage in 1832, and it was also the portage traversed by Morrison previous to that time, when he wintered at the eastern end of Fishhook lake, missing a meeting with Lieutenant Pike, which would have proved an historical event.

Continuing our voyage, evidences of an ancient occupancy were discovered in the valley of the Crow Wing, notably at the Eleventh lake and at the eastern extremity of Colonel Martin's Elbow lake. During the continuance of our voyage we gathered from the natives and others all possible information concerning the well-known imperishable signs of the Mound-Builders, which, coupled with our own observations, proves beyond question that prehistoric man migrated to and occupied the entire upper water-shed of the Mississippi, from Itasca lake to the mouth of Leech Lake river, and downward from there to the Sandy lake locality; that all the portages from lake to lake and from river to river, so extensively used even to the present time, were not discovered and opened by the Sioux or Ojibway Indians, but are prehistoric in character; and that the tribes named came after the Mound-Builders in the use of this entire system of portage communication. The mound-building population, whosoever they may have been, first traversed the Cut Foot Sioux portage, the portage from Beltrami's Julian source to Red lake, the portage from Pike bay to Leech lake, and the Shingobi portage, occupying for an unknown period of time the whole extent of territory drained by the upper branches of the Mississippi, residing usually upon the shores of lakes near the outlet or inlet, or both, in villages, subsisting principally upon game and fish, and using extensively pottery vessels made of pounded stone and clay, sand and clay, or pulverized mussel-shells and clay, stone implements, the bow and arrow, stone spears, copper implements, and either skin, bark or log canoes. The size of these people was undoubtedly from five feet six or eight inches, to six feet one or two inches, in height, as evidenced by the exhumed remains examined. The regular and symmetrical formation of the skulls examined indicates a high order of tact and sagacity on the part of this lost race; and it seems reasonable to presume, as I believe, that the effigy mounds near Tascodiac lake constituted a place of worship or celebration of some significance, for this mound-building people. That the flesh was removed from the bones previous to burial seems certain, but in what manner is doubtful; and the purpose was apparently to preserve the remaining bones by a process almost a hermetical sealing in character. I noticed no evidences of cannibalism. The war-arrow point, triangular and without notched base, seems to have been commonly used.

Whence came these people, and how and when did they depart? are questions that I do not believe can be correctly answered. Concerning the date of their occupancy of this remote and ultimate reservoir system of the Mississippi basin a final determination may be formulated from one of two propositions: First, that there may have been a large number of people there for a comparatively short period of time, or, second, a limited number for a much longer period. The preference would be for the latter suggestion, for a portion of that which remains after them appears of great age and of a remote antiquity; and scientists, to whose acute judgment I am perfectly willing to yield, will not surprise me in entertaining an opinion that the locality examined was occupied by a mound-building race of men more than twelve hundred years ago. All that I claim for the few months' labor I have devoted to this subject, entirely at my own personal expense (with the exception of the amount paid by the late Mr. Hill), can be embodied in a few words. It can now be correctly represented that a mound-building people formerly occupied the entire extent of the basin of the Mississippi from Itasca lake to the Falls of Pokegama; that the principal portages throughout that locality are the portages formerly discovered and opened at an unknown date by the same people, and that they were a race of men superior to the Ojibway population now occupying the locality, as evidenced by facts which have now come to light; and that those facts can be augmented very materially by a detailed survey and examination of the village sites, mounds and portages now known to exist there. There is no reason why statements should be accepted as true, unless there is first the most convincing and indisputable proof offered to substantiate a fact stated, and these facts now stated for the first time are to me as indisputable as they are interesting and instructive. Among the seventy-five or eighty remains exhumed, the Itasca and the Tascodiac skulls show a remarkable perfection of the human brain at that early period, as regular in symmetrical outlines and formation as the white population of the present time.